First Printing, 2014

ISBN-13: 978-1499625394
ISBN-10: 1499625391

Printed in the United States of America

Dedication

To the municipalities and operators whose NPDES licenses are on the line to be in compliance with EPA Regulations and the Clean Water Act laws...

To the neighborhoods that have lived long enough with the odors produced by the daily collection of sewage in a wet well lift station nearby...

To the city managers and / or the operations managers of the POWT's (Publicly Owned Waste Treatment) plants who are trying to control costs and stay within budget every day...

To the Wastewater Collection Managers who work hard to maintain the sewage lines and the lift stations, cleaning them of trash and debris (sometimes daily), and pull, unclog, and reset pumps...

There is a new technology in pumping wastewater and it will make life easier for the operators, the maintenance people, the city management, the users and the neighborhoods. It will cost less to operate, is self-cleaning with less to no clogs, and can be operated and managed from an office, a laptop, or a smart phone.

It is called The DIP System®

and it is revealed in detail here.

Table of Contents

Introduction

The DIP System® was developed by S.I.D.E. Industrie in France near Paris by Stéphane Dumonceaux. S.I.D.E. Industrie is a family owned company that has specialized for over 25 years in solutions for the pumping of "difficult fluids."

Based on practical expertise in the field, the development of the DIP product range is the result of 30 years of research, and from listening to the daily concerns of users and designers, enabling S.I.D.E. to offer today the very innovative solutions which are both simple and complete. Many of the systems are already the standards in use in Europe and have been for a number of years.

This product, new to the United States, presents innovations and new development designed to protect the environment as well as the investment in the equipment.

Each system is individually manufactured and tested before it is shipped for an almost "plug and play" installation. It is assembled and shipped from Kansas.

Once installed, operator and maintenance personnel will immediately benefit from the safety of no longer having to deal with the dangers of hydrogen sulfide (H_2S), the odors, and can do any service work in a clean, dry environment.

The DIP System® is patented and is proven with years of use.

Chapter 1

The history of wastewater management
(From Wikipedia)

Sanitation in ancient Rome was a complex system similar in many ways to modern sanitation systems. During the Dark Ages, the technical knowledge of the Roman system was lost and has subsequently been investigated by modern-era historians and archeologists.

The latrines are the best-preserved feature at House Steads Roman fort on Hadrian's Wall. The soldiers sat on wooden boards with holes, which covered the big trenches. Water ran in the two small ditches at the soldiers' feet.

A system of thirteen Roman aqueducts provided the inhabitants of Rome with water of varying quality, the best being reserved for potable supplies. Poorer-quality water was used in public baths and in latrines, which were an early form of toilet. Latrine systems have been found in many places, such as House Steads, a Roman fort on Hadrian's Wall, in Pompeii, Herculaneum, and elsewhere that flushed waste away with a stream of water. Romans used sea sponges on a stick & dipped in vinegar for wiping after defecation.

The Romans had a complex system of sewers covered by stones, much like modern sewers. Waste flushed from the latrines flowed through a central channel into the main sewage system and then into a nearby river or stream. However, it was not uncommon for Romans to throw waste out of windows into the streets, at least according to Roman satirists. Despite this, Roman waste management is admired for its innovation.

It is estimated that the first sewers of ancient Rome were built between 800 and 735 BC. Drainage systems evolved slowly, and began primarily as a means to drain marshes and storm runoff. The sewers were mainly for the removal of surface drainage and underground water. The sewage system as a whole did not really take off until the arrival of the Cloaca Maxima, an open channel that was later covered, and one of the best-known sanitation artifacts of the ancient world. Most sources believe it was built during the reign of the three Etruscan kings in the sixth century BC. This "greatest sewer" of Rome was originally built to drain the low-lying land around the Forum. It is not known how effective the sewers were, especially in removing excrement.

From very early times the Romans, in imitation of the Etruscans, built underground channels to drain rainwater that might otherwise wash away precious top-soil, used ditches to drain swamps (such as the Pontine marshes), and dug subterranean channels to drain marshy areas. Over time, the Romans expanded the network of sewers that ran through the city and linked most of them, including some drains, to the Cloaca Maxima, which emptied into the Tiber River. In 33 BC, under the emperor Augustus, the Cloaca Maxima was enclosed, creating a large tunnel. The Cloaca Maxima, it is presumed, built in the fourth century BC and reconstructed under Augustus), still drains the Forum Romanum and surrounding hills. Strabo, a Greek author who lived from about 60 BC to AD 24, admired the ingenuity of the Romans in his Geographica, writing:

"The sewers, covered with a vault of tightly fitted stones, have room in some places for hay wagons to drive through them. And the quantity of water brought into the city by aqueducts is so great that rivers, as it were, flow through the city and the sewers; almost every house has water tanks, and service pipes, and plentiful streams of water"...In short, the ancient Romans gave little thought to the beauty of Rome because they were occupied with other, greater and more necessary matters.

A law was eventually passed to protect innocent bystanders from assault by wastes thrown into the street. The violator was forced to pay damages to whomever his waste hit, if that person sustained an injury. This law was enforced only in the daytime; it is presumed because one then lacked the excuse of darkness for injuring another by careless waste disposal.

Around AD 100, direct connections of homes to sewers began, and the Romans completed most of the sewer system infrastructure. Sewers were laid throughout the city, serving public and some private latrines, and also served as dumping grounds for homes not directly connected to a sewer. It was mostly the wealthy whose homes were connected to the sewers, through outlets that ran under an extension of the latrine.

In general, the poor used pots that they were supposed to empty into the sewer, or visited public latrines. Public latrines date back to the 2nd century BC and whether intentionally or not, they became places to socialize.

Long bench-like seats with keyhole-shaped openings cut in rows offered little privacy. Some latrines were free, for others small charges were made.

According to Lord Amulree, the site where Julius Caesar was assassinated, the Hall of Curia in the Theatre of Pompey was turned into a public latrine because of the dishonor it had witnessed. The sewer system, like a little stream or river, ran beneath it, carrying the wastes away to the Cloaca Maxima.

The Romans recycled public bath waste water by using it as part of the flow that flushed the latrines. Terra cotta piping was used in the plumbing that carried waste water from homes. The Romans were the first to seal pipes in concrete to resist the high water pressures developed in siphons and elsewhere. Beginning around the 5th century BC, city officials called Aediles supervised the sanitary systems. They were responsible for the efficiency of the drainage and sewage systems, the cleansing and paving of the streets, prevention of foul smells, and general oversight of brothels, taverns, baths, and other water supplies.

In the first century AD, the Roman sewage system was very efficient. In his Natural History, Pliny remarked that of all the things Romans had accomplished, the sewers were "the most noteworthy things of all".

The system in Rome was copied in all provincial towns and cities of the Roman Empire, and even down to villas that could afford the plumbing. Roman citizens came to expect high standards of hygiene, and the army was also well provided with latrines and bath houses, or thermae.

Roman rubbish was often left to collect in alleys between buildings in the poor districts of the city. It sometimes became so thick that stepping stones were needed. "Unfortunately its functions did not include house-to-house garbage collection, and this led to indiscriminate refuse dumping, even to the heedless tossing of trash from windows." As a consequence, the street level in the city rose, as new buildings were constructed on top of rubble and rubbish.

The History of Plumbing- Pompeii and Herculaneum

The ancient Roman towns-cities Pompeii and Herculaneum were being destroyed in 79 AD when Mount Vesuvius erupted. The volcanic pyroclastic flows buried the territories of the Pompeii and Herculaneum communes under 13 to 20 feet of vast pumice and ashes.

The excavation of the Archeologists in Pompeii and Herculaneum, which started in 1758, uncovered many ancient Roman buildings all throughout the land of the Roman empire. Relics of walls, aqueducts, palaces, forts, theaters, hospitals, ordinary homes and temples were also found. These excavated cities have helped us in better understanding the Romans' way of life during ancient times.

Romans were not only known for being the pioneers of the new building designs and materials but also for their avant-garde water channels. They were also known to be great water engineers. They had devised a great number of raised channels to bring water from the water sources of distant mountains and hills into towns and cities. They had also invented water pumps with valves. The Roman valve helped in pumping water supply uphill. The pumped water then went into the high tanks that were located above the fountains. From the tank, the water then came out of the spout through gravitational pull.

Overview of the Pompeian Plumbing

As early as ancient times, Pompeii plumbers already had a booming profession. They were always busy as there were constant demand for pipes, bronze valves and wiped joints. There were also steady orders for gold, silver and marble opulent fixtures. Among their major works was molding and fabricating lead gutters for the richest Pompeian private homes. The Pompeian homes, in general, were built with open-roof design and an atrium. There was usually a tank on the base part of the house which was used to collect the rainwater that had fallen down from the tiles of the roof.

Pompeii plumbers fashioned the pipes by pouring molten lead into sheets with various dimensions and thickness. The liquefied lead was then exposed to open-air to let it cool. When the sheets were cool enough, plumbers then formed the sheets around a wooden core. A V shaped opening was also left so that the ends of each of the now molded pipes will meet and fit each other. The plumbers then formed a clay or sand mold all around the channel. Once molding was done, the hot lead was then be poured into the opening to make the joint. The typical pipes during ancient times were elliptical or egg- shaped.

During early Roman times, only the richest Pompeian homes were able to get constant supply of running water. Each of the private homes paid for their water supply based on the size of the nozzle they used. Those ordinary people, on the other hand, needed to drink and fetch water from the public fountains. The water was being carried to different homes through the lead pipes. The utilization of the lead pipes had not only helped in the carriage of water but also greatly improved the sanitary conditions of the city.

The Romans' ancient plumbing systems were among the most advanced in the world and were considered a huge part of the city structures. In fact, their water supply systems were so advanced that no one had anything else better until 1800s. The invention of the Roman aqueducts also paved way in the development of underground sewer systems. The largest known sewer that was built in Rome during ancient times was the "Cloaca Maxima". It was wide and high enough that even a horse with its cart could be able to drive through it.

Out of the development of the ancient Roman water supply systems, the Roman engineers were also able to construct public lavatories. Although these lavatories were convenient enough, they were never private. These were designed in a way where the users had to sit side by side on the arranged rows of seats.

Ancient Pompeian Private Bathroom

The bath room of the richest Pompeian private homes was literally a room having a pool. Generally, the entire floor of the room was filled up with water. It modern times, this is called a small sized swimming pool. Marble was primarily used as the wall lining. There were also three to four complementary marble steps which served as the stair case that connected the surface area and the submerged concrete floor.

The whole room temperature was regulated by a special kind of damper system. Both the water and the air that circulated around the room were heated to achieve the desirable temperature so as to add comfort to the bath users. The hot air was drawn from the adjacent furnace on the piers of bricks of which the entire floor rested. There were also terra cotta tiles that were used to line all the sides of the walls so the heat could easily pass through.

There was also a built-in plug in the bath. This would allow the users to empty the bath. Normally, the bath was emptied and refilled once or twice a day depending on the usage. The pipes used were also built to last. They may be fashioned with lead or perhaps with tiles buried in the ground. When tiles were used, they were set no less than a foot deep. The buried tiles were then covered with a very solid concrete floor.

Roman Public Baths

The public baths in Rome normally had silver valves. Those which included brass faucets wiped joints, four-branch crosses or fittings and single-sized bronze bathtubs were considered luxury plumbing. There were also separate chambers intended for dressing and undressing, cold water baths, tepid baths and steam baths.

According to the archeologists, they had excavated a sort of bath complex in the Pompeii area. This complex was said to measure almost a mile in perimeter- big enough to be considered as the second largest complex next to the amphitheaters.

Accordingly, the Herculaneum catered to only the most powerful and the richest among the rich families. It was a resort that was very expensive, in fact, more costly than the events in Pompeii. The exquisitely designed mosaic in the bath of the women as well as the murals on the walls were said to be superior to those of Pompeii.

Initially, the concept of the Roman public bath in the ancient times was generally comparable to the country club in the modern generation. It was a special place where people socialized. It was where they meet their friends, enjoy playing games, eat a great meal and exercise in the gym. It was also in the public bath where people spend a bit of their luxury time enjoying a series of relaxation baths; warm, hot, cold and tepid baths.

Romans during Julius Caesar time in 100-44 B.C were conservative, especially with the issue of gender mixture. During this time, women and men were not allowed to bathe in the same room. There was separate bath time for men and a separate bath for women. Each of these separate baths were set for a couple of hours. Romans usually bathed stark-naked. Women, however, use a cap to secure their hair and their pearl necklaces.

In later times, this rule was modified. Both men and women were then allowed to go in one bath at the same time. Both sexes also bathed without their clothes on. There were just two main new rules added this time. The first rule was "don't stare" and the second one was "act as if you were dressed up". These two rules were strictly implemented. Those who violated the rules were asked to leave the bath immediately and were then not allowed to enter the bath again.

Roman spas during the ancient times were always associated with luxurious lifestyle. This lifestyle had overwhelmed even the Romans themselves. As Rome degenerated, the baths likewise degraded over time and became a place for debauchery and immorality.

Chapter 2

Where are we today?

Two thousand years and not much has changed in the management of wastewater!

We are still using terra-cotta tile, and pipes (and still some old brick tunnels in some older cities) now replace the stone covered waterways. Gravity still works! Sewage still flows to some lower point to be collected. They are now called wet wells and in rural areas, septic tanks. The odors are still the same as then. The septic systems use tiles to disperse the "grey water" into the surrounding landscape. The "solids" settle out and this is what is pumped from septic tanks as sludge.

The wet wells, now called "lift stations," are usually associated with high density neighborhoods as they collect and retain higher volumes of raw sewage. They are designed to turn on a pump when a certain level is collected. The pump transfers the sewage to a treatment plant or to another lift station to be collected and transferred again, ending at a treatment plant..

Now, rather than emptying the raw sewage into a river, the Clean Water Act requires sewage be treated and meet high standards before being discharged.

As wet wells are filled by anything that flushes or is forced into the sewer line, they are notorious for collecting all sorts of trash and debris. Rags, diapers, plastic bottles, cans, grease, oil, dead animals, and if near, prison uniforms, all of which need to be physically removed and hauled away. These things also clog the pumps and cause maintenance personnel the smelly job of removing the clogged pump, cleaning it and then replacing it. (Time and time again!)

What if you could design a lift station that?

- Automatically follows any variations in flow
- Automatically adjusts its head pressure to the variable flows
- Handles trash and debris with no clogging and is self-cleaning
- Accepts entrained air up to 10% without cavitation
- Even handles dry running
- Reduces or eliminates H_2S (No gas and no odors)
- No sand or grease accumulation
- No need for screens, rakes, or regular cleaning of trash or debris accumulated in the wet well
- Eliminates the need for wet wells altogether
- Presents a clean, safe environment for service personnel for minimum routine maintenance
- Reduces energy costs
- Costs less to maintain and install
- Communicates and is remotely serviced and programed

Now comes the DIP System®!

Chapter 3

What is the DIP System®?

DIRECT IN-LINE PUMP SYSTEM

IT IS AN INNOVATIVE PRINCIPLE

By lifting gravity effluent flow directly at the point of entry, without water loading or a wet well, the **DIP System®** overcomes the drawbacks of retained wet well volumes of effluent as there are:
• No dangerous gases (H_2S),
• No odors,
• No sand and grease accumulation,
• No equipment corrosion,
• No structural erosion,
• No clogged float switches,
• No trash or debris collection,

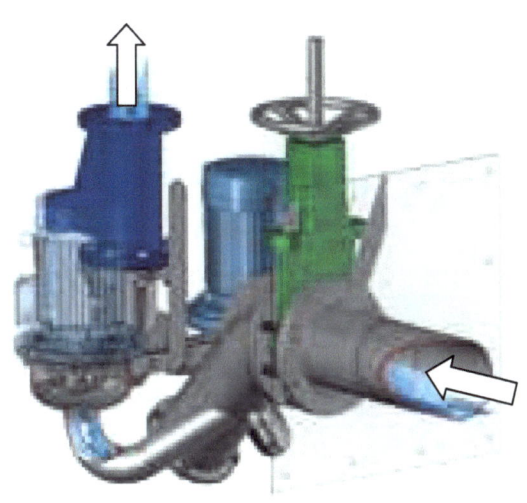

What you <u>now</u> <u>have</u> with wet wells

A wet well with accumulations requiring cleaning.

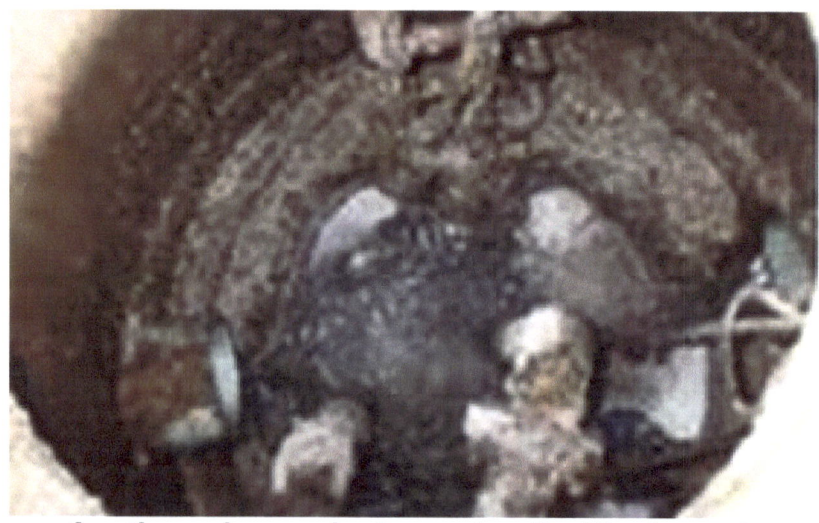

Another odor producing wet well with clogged pumps.

With the DIP System® (seen here)

- **No Wet Well Trash Removal and all Associated Costs with Disposal**
- **Reduced Energy Costs**
- **Reduced Manpower Costs Needed to Remove and Reinstall Clogged Pumps**
- **Eliminated Odors at Site and / or in Neighborhoods**
- **Eliminated Need for Screens and Rakes**
- **Eliminated or Reduced Lift Station Cleaning Requirements**
- **Eliminated Personnel Exposure to H_2S (Hydrogen Sulfide)**
- **Eliminated Need for Wet Well Use Totally**
- **Less excavation needed for new construction**

You _can_ convert your existing wet well to a DIP System®!

The DIP System® makes it possible to design durable and economical lift stations.

IT IS A COMPLETE CONCEPT

A main sewer knife valve is supplied as standard for shutting off the wastewater inlet.

CONICAL VORTEX IMPELLER

This impeller is fitted as standard as is perfectly adapted to raw effluent full of air and sand. **The special properties of these alternate blade impellers allow most fibrous and solid materials such as cloths, bandages, clothing, plastic bottles, aluminium cans, etc. to pass through without causing clogging**. It prevents stoppages and helps the system to reprime quickly. This impeller is the best compromise between unclogging, sustainability and efficiency. Conical **vortex** impellers are not affected when dry, and can run dry without causing damage, for a period of several weeks.

T4 IMPELLER

Multicanal open impeller made of AISI 304L or 316L stainless steel has a wide flow section offering an optimal efficiency on water flow. The T4 version is fitted to the larger DIP System models and has a high level of efficiency. Coupled to the new IE3 Premium motor this impeller allows an operation to save 30% of the energy.

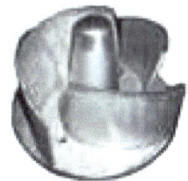

DIPCUT® IMPELLER

Another new innovation:
DIPCUT® is a conical vortex impeller that becomes a "Shredder" when it automatically changes direction of rotation when coupled to a smart automated command. This impeller changes its direction of rotation when needed in order to cut snarled long fibrous materials and rags, and then removes them. This is the ideal "Rag Killer!" DIPCUT® is the perfect combination between the advantages of the conical vortex impeller pumping sand, air or big solid wastes and the shredding function cutting long fibrous materials into shreds. Cleaning out a classical pumping system costs an average $480. That is equivalent to an electric consumption of 3 MegaWatts. DIPCUT® allows the savings of pulling, cleaning, and resetting those clogged pumps and uses less power.

DIP System® IMPELLERS CAN BE COMPLETELY CHANGED IN ABOUT 15 MINUTES!

Contrary to the Grinder or shears that cut and pump, the DIPCUT® impeller keeps its high hydraulic pumping efficiency. Moreover, while shredding, the power of the motor is used only by the 4 "knives" that use much less energy resulting in greater efficiency. This efficiency is half of that when compared to a grinder system having the need for larger motors.

The switching between the 2 functions, pumping and shredding, is managed by a signal based on the torque control and is monitored via the OmniDIP® Control Box connected into the command module. Additionally, the self-monitoring OmniDIP® (See Chapter 10) allows the remote follow up and analyses of the function of the DIPCUT®.

Two electric motors are connected by a hydraulic body, the shapes of which are specially designed to be able to receive effluent directly. The upstream level is measured by a static sensor, fitted in the water stream of the effluent inlet.

All parts in contact with fluids are made of "boiler-plated" stainless steel EN1.4306 or EN1.4404 (304L or 316L).

Chapter 4

SHARED HYDRAULIC BODY

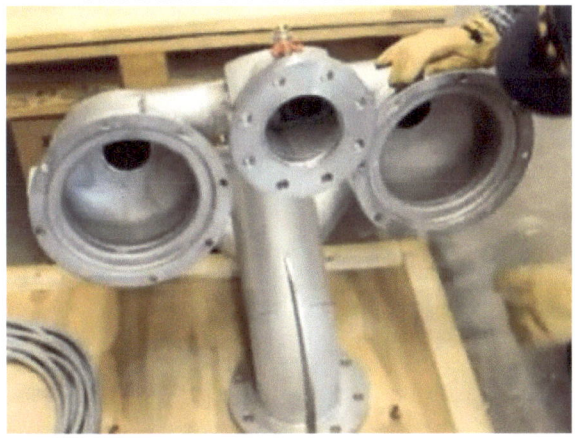

Made entirely from AISI 304L (316L on request) "boiler-plated" stainless steel. The bodies' suction profiles are specially designed to take advantage of the flow speed from the gravity-driven inlet. The inlet body also includes a "stone trap" with an inspection port and draining valve.

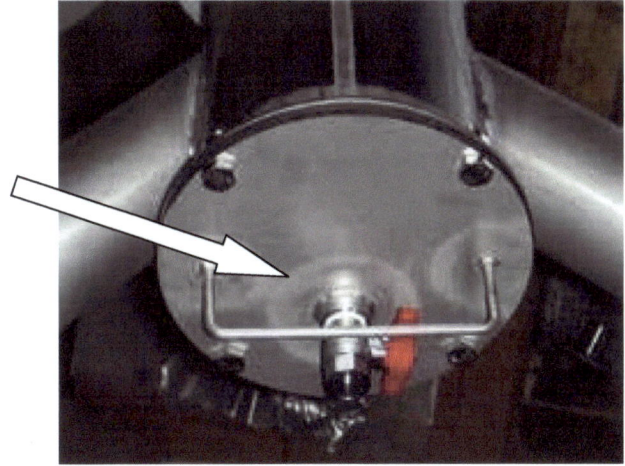

25

The interior surface of the pumping system body is very smooth to improve efficiency, and has no areas where debris in suspension might be collected. The wide flow section continues through to the internal directional swing check valve.

The swing check box is an integral part of the body, eliminating the need for additional pipework between the two pumps. The valve has three working positions: right or left according to which pump is operating, and central if both are operating. It has a stainless steel frame and replaceable wear plates.

The joined discharge, 3" to 16" is sized to the model based on the flow requirement, has a Standard ANSI compliant flange and includes a socket to install a pressure measuring device. A single check valve must be fitted directly to this flange to control the volume contained within the discharge pipe, typically a rubber swing check valve.

Chapter 5

MOTORS and SEALS

As standard: Cast iron or aluminium motor body with NEMA fastener dimensions which comply with industrial standards and are, therefore, compatible with those of regular industrial motors. Motors are TEFC, Class F windings (310°F). Efficiency class IE2. Sealing systems are, at level IP56 meeting, the same demands as for running machinery installed in the holds of ships.

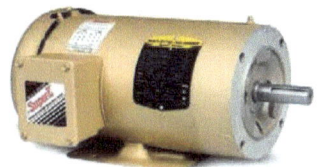

"Immersible" version:
On request, we can supply motors with IP67 "immersible" protection fitted with armored wiring (length to be determined upon request) and a resin sealed connection.Efficiency class IE3, Meeting NEMA Standards, these all-stainless steel motors and their encapsulated stator are designed to operate in all environments and to withstand all moisture attacks. They are ventilated and immersible under 25' of water for 2 weeks. (Typically for locations pumping sewage or industrial fluids located in a flood zone)

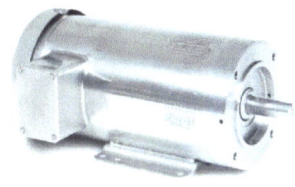

SEALS

- No oil reservoir needed
- Ability to operate dry up to 150 hours
- Large sealing capacity
- Also operates under pressure
- Materials: Sic/Sic or W/Sic

OPERATING LIMITS

- Max. Temperature: 100°F (180°F on request)
- Viscosity: 750 cSt (ie 1.08×10-5 ft²/s)
- Max. Speed: 3,600 rpm
- Max. Pressure: 145 psi

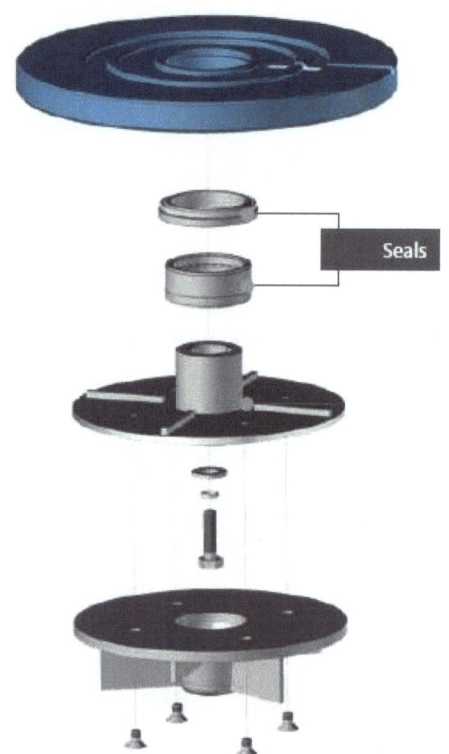

IMPELLERS CAN BE CHANGED
IN AS LITTLE AS 15 MINUTES!

LEVEL GAUGE SENSOR

A pressure sensor located beneath the effluent entry chamber constantly measures the height of fluid at the inlet.
As it is stainless steel AISI 316 with a flush membrane, this sensor is highly wear resistant. It is also resistant to deposit build-up because at this position, it benefits from the inlet fluid speed, which is further enhanced by the suction effect of the pump operation.

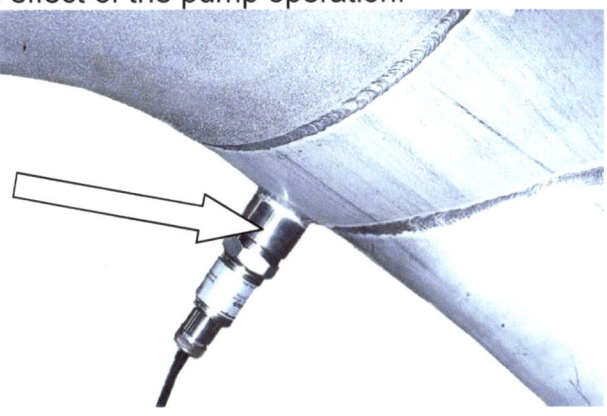

This system dispenses with the need for classical detection methods, such as float switches or ultrasound measurement.
IP67 protection with 50' of cable is standard on all models.

In addition to the information transmitted to the control panel regarding the inlet fluid height, information from the sensor can also be sent by a transmitter for remote management of the system, without the need for any accessories. It comes with a dry contact output on the cotrol panel.

Chapter 6

OPERATION

The **DIP System**® is driven, as standard, by variable speed. Operation is no longer based on "all-or-nothing" or a fill the wet well and pump down in a batch. The **DIP System**® now operates continuously and modulates pumping directly from the effluent inlet.

Thanks to its special design, the **DIP System**® absorbs the air/fluid mix which flows in from the gravity lines and gives it the speed it requires to be discharged up to the outlet. The proportion of air transported can reach up to 10% of fluid flow without running the risk of cavitation.

Flow is also maintained by the system, which automatically adapts to constantly changing flow rate and load reduction, due to the special shape of the hydraulic body and the design of its impellers.

This operating mode enables solid or fibrous matter to move through the system without causing blockages. Electrical power adjusts in line with the incoming flow. Progressive start-ups and stops eliminate hydraulic surges. If the inlet flow is less than the minimum load, operation is intermittent and, if the flow is zero, the **DIP System**® shuts down completely until flow resumes. Again, the internal swing check valve provides integrated flow without the need for complicated piping systems.

Chapter 7

CONTROL PANEL

The DIP System® comes delivered either with its ALC (Advanced Level Control) panel to be assembled in a customer provided enclosure, or in a complete cabinet.

Each Variable Frequency converter is connected to its motor unit and communicates with the other. Speed variation and simplified control levels on the same panel allow regulation in all configurations, including those of complex combined sewage systems.

The ALC panel can be used to carry out all the functions of a traditional lifting station without the need for additional equipment:
• Automatic alternation,
• Emergency stop,
• Automatic backup,
• Manual control,
• Automatic cascade,
• Automatic rotation direction reversal for clearing,
• Auto-setting of operational limits.

CONTROL PANEL (Continued)

The control system also provides a very high level of integrated protection for:
• Over intensities, over voltages and under voltages
• Sensor faults
• Internal faults
• Grounding faults
• Auto-diagnostic
• Impeller blockage
• Fault log
• Phase loss
• Emergency stop
• Phase direction
• Remote communication
VIA MODBUS
• FACTORY remote maintenance VIA GPRS
• 2 x DIP controls in tandem or in parallel

FLOW REGULATION, even when highly variable:

The DIP Systeme® automatically adapts to the incoming flow, up to the limit of the total flow of 2 motor maximums, i.e. from 0 to 200% of the nominal flow. The performance of a DIP model is between 2 and 4 times higher than the flow rate achieved by traditional pumping in batch mode.

HUMAN-MACHINE DIALOGUE

On the front, an Auto/0/Manu switch and a continuous display of 3 key pieces of information. The control panels are removable for safe keeping. There is a simplified display for easy use:
• Values displayed for: speed, intensity, level gauge, motor power, motor torque, and meters.
• Status readings for remote user management.

PROVIDES CONSTANT AND REGULAR FLOW

Located upstream from the treatment plant, the DIP System® provides constant and regular flow. It avoids fluids arriving in "batches", usually detrimental to the biomass used for biological treatment. The system management also limits maximum outlet flows.

REDUCTION OF WATER HAMMER

The DIP System® uses a soft start ramp on start-up and a deceleration ramp before stopping each pumping unit to eliminate valve shocks. During diphasic pumping (liquid + air), water hammer can also be reduced.

ENERGY SAVINGS

The problem of reducing the number of start-ups no longer exists and energy savings can be realized at low flow rates. The delivered power for the motors is automatically adjusted in line with the required flow rates.

OmniDIP®: REMOTE CONTROL

OmniDIP® is a SCADA system with remote control and management in mind, and is based on M2M communication dedicated for DIP System®. It allows many advanced remote functions as reseting, remote unclogging and information via secured Internet interface and/or standards SCADA controllers.

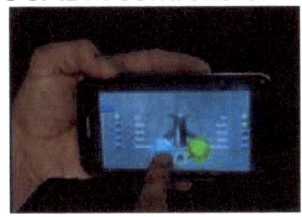

Chapter 8

ENGINEERING FOR NEW OR EXISTING

As part of a lifting station installation project, the **DIP System®** enables civil engineering costs to be significantly reduced:

- Sitework requires, at least, 3 feet less in foundation depth and concrete volume.
- Takes up less space than a traditional station.
- Reduces the depth of the structure to 20" below intake level
- The structure's shape can be either round or square and commercially available ducting and pipes prefabricated in concrete are more than adequate to contain the DIP equipment and the valve systems.
- Dry installation enables the single check valve and gate valve to be assembled in the same location as the **DIP System®** without the need for a separate valve chamber.
- A single inspection port is all that is required.
- For inlet levels which are not very deep, the equipment room can even be constructed out of concrete blocks as there is no "pit" to be flooded, and it is typically watertight.

During station renovation: The **DIP System®** adapts to any type of currently available pipework; precise positioning of input/output pipes is no longer required. The discharge head can be positioned at any angle of direction through 360°.

ABSENCE OF WET WELL

The now <u>dry</u> wet well or dry holding tank becomes an equipment room which can be fitted with lighting, a ladder and other accessories which enable maintenance personnel to carry out their work in clean, dry safety.

The lifting station becomes a straightforward inspection chamber without human danger (no emission of dangerous gases, odors or accumulation of solid matter).

The **DIP System®** equipment is rustproof, and is thus more resistant and more durable.

SIMPLIFIED MAINTENANCE

The absence of a wet well eliminates regular and costly cleaning operations of traditional units.

INCREASES THE CAPACITY OF EXISTING INSTALLATIONS

The **DIP System®** enables flow or discharge capacities to be increased in a pumping station which has insufficient power, *without* changing the civil engineering.

CONSTRUCTION RULES FOR THE DESIGN OF DIP STATION STRUCTURES ARE AS FOLLOWS:

Ideally bring the various inlets together in an inspection chamber before the station, at least 15 ft from the DIP. The slope between the chamber and the station must be equal to or greater than 2%. New units can have a lesser slope, for existing or new installations. However, the inlets may be brought together using a set of wall flanges in the DIP enclosure.

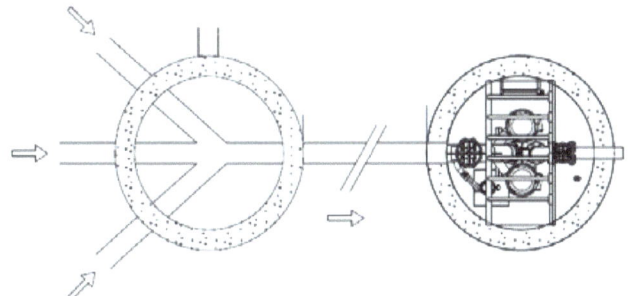

The space between the inlet and the floor must be a minimum of 20 inches. This height is required for easy access of the sensor (See dimensions according to the model selected).

Apart from the minimum space around the motor units needed for maintenance, there are no constraints in terms of shape or volume for the enclosure containing the DIP System®. (The dimensions of each model are supplied with the technical data sheets upon request.) The form of the structure and the diameter of the inlet pipe must be specified when ordering the wall flange. The top of the wall flange is fitted with a return tap for drainage from the pump discharge sump.

The construction may be of any type: the **DIP System®** can be fitted into a round or square concrete structure, a simple inspection chamber made of concrete ducting stacks or an enclosure made of concrete blocks for a collection network that is close to the surface.

The discharge direction can be chosen through 360° as the guiding system has no special requirements and the discharge pipe is shared by two motor units.

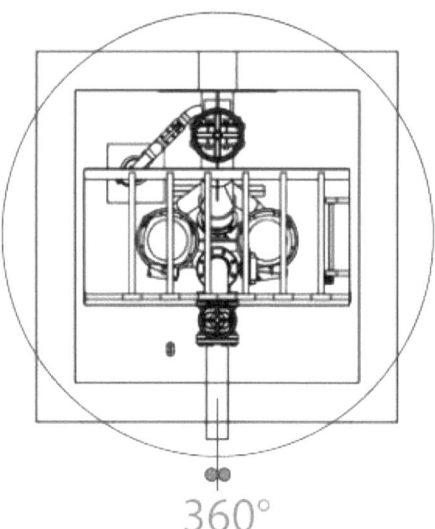

360°

Discharge only requires a gate and a check valve, except when it is a lifting only station. All the valve equipment and any measuring devices (flowmeter, sampling points, etc.) can be fitted in the station without separate valve chambers, as the enclosure area remains clean and dry.

For deep enclosures fitted with an intermediate floor, sufficient ventilation must be provided.

A double trap door 28" x 56" is recommended up to the DIP101 models. Subsequent models require a double trap door directly above the system, of suitable dimensions for the size of the model selected.

DIFFERENT POSSIBLE SHAPES OF UNITS: ROUND, SQUARE, RECTANGULAR...

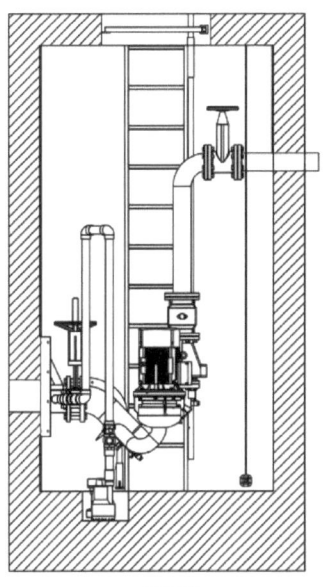

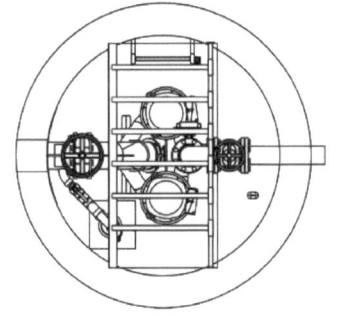
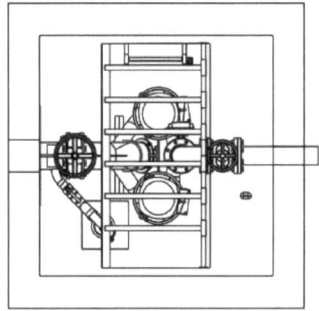

MOTOR CONNECTIONS

Motor connections should preferably be made without a break between the variable frequency drives and the motors. Preferred cables are LIYCY armored cables. Connect to the motor terminal boards taking into account the voltages shown on the ratings plates.
IP56 protection of standard motors requires careful assembly of the compression fittings. IP57S versions are equipped with 30ft of shielded lead cable per motor as standard.

GAUGE WIRE SELECTION
(STANDARD VERSION) for 460V

(Maximum length of motor cables for power in excess of 2.5HP = 990 feet, and 330 feet for power below 2.5HP).

POWER	INTENSITY	WIRE
2 to 4 HP	3 to 7 A	16 AWG
5.5 to 7.5 HP	8 to 12 A	14 AWG
10 to 15 HP	15 to 19 A	12 AWG
15 to 20 HP	21 to 28 A	10 AWG
25 to 30 HP	34 to 42 A	8 AWG
40 to 50 HP	55 to 67 A	6 AWG
60 to 75 HP	90 to 110 A	4 AWG
100 HP	150 A	2 AWG
120 to 150 HP	180 to 220 A	1/0 AWG

44

CONNECTION OF IP67 SENSOR

If possible, the sensor should be connected without a break between the cabinet and the sensor.
The delivered gauge wire is armored to protect the signal. It does not matter if the sensor is initially connected incorrectly as the display is protected. The motor cables used are armored; ensure a wire routing distance of at least 2 inches is maintained between them and the sensor cable.
If cables cross, place it on top of the power cables at an angle of 90°.

CONTROL SYSTEM

The cabinets and the control panels are delivered ready to be plugged in.
Electrical connections should be carried out by qualified personnel.
Only the two motors and the sensor require connection for operating the DIP. All internal connections are made and tested in the factory. Dry contacts for remote management status reports are installed as standard. Refer to installation and maintenance instructions.
The required general protection is of the 300 mA differential interrupter type, except for special requests.

For single panels to be fitted into a customer supplied enclosure, correctly calibrated motor starter protection must also be fitted upstream of each Variable Frequency Drive (magnetic circuit breaker or fuse holders for example).

COUPLED UNITS

For wide variations of input flow, unit variants exist enabling two DIPs to be connected in parallel. They can also be connected in series for higher elevations than standard, up to 300ft.

PERFORMANCE RANGES

The performance range of each model shows the automatic adaptation zone of the flow/height ratio.
The series of DIP 21 - 31 - 61 - 101 – 131 - 151 etc. allows contraction of the column of water at the DIP inlet, which is less than 10% of the nominal flow. Accordingly, gentle slopes or flow variations beyond a range of 0 to 200% can be dealt with. Minimum load curves therefore no longer represent the lower flow limit on these new models.

Chapter 9

MODEL SELECTION

The peak input flow rate and the corresponding total pressure directly determine the DIP model to be selected. In-line pumping does not require additional coefficients to be calculated (number of start-ups or drawdown volume): the DIP directly adjusts itself in line with the inlet volume and can switch its motors on and off up to 150 times an hour without harm.
The maximum inlet flow must correspond to one of the points on the upper curve of the operating range graph (supplied upon request). The graph will show the performance with a single motor unit in service, maintaining *full 100% backup* with the second motor unit. For all the lower operating points, the system automatically adjusts its flow and power at maximum efficiency for the total pressure head to be overcome.

FUTURE EXPANSION?

These units can be supplied in variable flow from 0 to 400% and linking networks, etc. In the event of extensions being planned to the collection network, you can choose the corresponding model for the maximum future output with the knowledge that from the start, the system can operate constantly at the bottom of its range without excessive consumption. e.g. A DIP31 with rated power of 2 X 4 HP used at 30% of its operating range will actually consume 30% of the power of only one of its motors, i.e., 1.2 HP.

Chapter 10

The Optional OmniDIP®
Remote Management System

Lift stations are naturally isolated, and usually far away from the treatment plant and management. However, they must assure influent transfer 24 hours a day with the maximum of safety and the minimum of maintenance. The **DIP System®**, direct in-line pumping is already providing a modern solution to this need:

- no wet well
- no smells, nor dangerous gases
- no operational cleaning
- no float switches
- no clogging
- can run dry
- modulated and soft pumping
- working comfort and safe access for employees
- soft starts and stops
- automatically adapts to variable flow rates

The Self-Monitoring OmniDIP® complements the DIP solution by:

- Automatically monitoring the internal pumping operational data 24 hours a day
- Automatic analysis of the functions of the pumps
- Monitoring the general performance of the pumping system

49

- Secure collection of the data for each pump by the factory
- Rolling archival of the last 3 months of operation
- Optimizes all of the operational parameters while running
- Provides preventive maintenance and management reports

OmniDIP® is a Self-monitoring system dedicated to the DIP System® installation.
It automatically checks and continuously monitors all of the processes through 230 parameters per pump in order to guarantee the optimal operation. This avoids any useless, unnecessary service or maintenance visits out to the lift station. It analyzes so precisely that it allows management to **forecast and optimize services.** It does much more than only inform when there is a technical fault or to just log data. The factory service is remotely checking or updating the system via **OmniDIP®** and will automatically handle pumping issues. The factory will also provide preventive maintenance alerts.

Some of the automatic processes of the **DIP system®**, such as automatic clearing, automatic cleaning or adjustment of the level set point for example, can be done from the **OmniDIP®** remotely. We can know the state of the sensor, the state of the motor or we can test the automatic alternating for example owing to the expanded assessment that evaluates every second of operation while under the factory control.

If anything peculiar is detected a summary report is sent to the user. In the event a site visit is required, our technician will contact and guide the user.
The Operator / User Access will allow a full display of the functioning of **DIP system®,** and the ability to edit summary reports over a chosen period of time. The user can directly check on the average inflow, the pumped volumes, the run times and the improvement actions made by the self-monitoring **OmniDIP®**.

The user can display synchronous data (such as torque, flow, frequency, current, upstream level for both pumps simultaneously) for a specific curve point. Automatic alerts can be sent either by mobile text message or email.

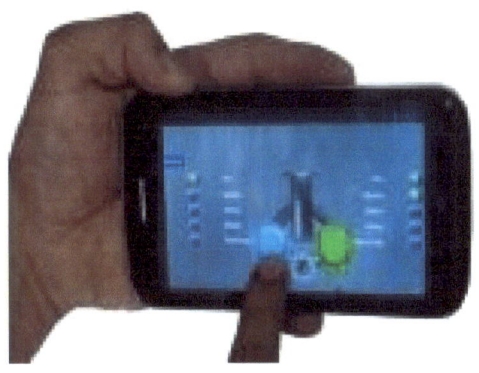

Conclusion

WHAT IS DIRECT IN-LINE PUMPING?
WHAT ARE THE ADVANTAGES?
WHAT ARE THE DISADVANTAGES?

The principle of direct in-line pumping is to pump directly from the gravity flow according to variations in water stream flow.

- This eliminates all the drawbacks of a wet well: dangerous gases (H_2S), odors, sand and grease accumulation, equipment corrosion, structural erosion and clogged float switches.

This process reduces costs in a number of different ways:

- 100% savings on regular operational cleaning costs.
- Reduced construction and investment costs, with a shallower pit and often less surface area taken up.
- Reduced maintenance costs as equipment is kept dry and not in contact with polluted effluent and gases.
- Ease of access, and designed with sustainable materials and few moving parts.
- With no oil reservoir and ability to run 150 hours when dry, the **DIP System**® is very forgiving of incorrect adjustments and its wearing parts are inexpensive.
- Reduced safety risk for operating personnel due to shallower structures and equipment that operates in a clean, dry environment.

DISADVANTAGES OF DIRECT IN-LINE PUMPING?

Just one! Sand, present in all networks in greater or lesser concentration, is also pumped directly, unlike immersed systems within which it decants. Result: The sealed system has been continually redesigned and improved to overcome this drawback, at low replacement cost and without requiring the external spraying normally required for mechanical parts, which are unable to operate dry. In the early years, this "disadvantage" was not insignificant, **but today it is no longer an issue**.

HOW MUCH DIRECT IN-LINE PUMPING CAPACITY DO YOU REQUIRE?

We are available to answer any questions you may have and to help you draw up an Equipment Selection Worksheet. (Contacts; Page 61)

For combined or rainwater networks, it is assessed using peak flow as with other pumping systems, You will have two motor units in service to avoid the oversizing that is all too often the cause of operational problems.

AND IF THE SYSTEM IS IN CONTINUOUS OPERATION? OR IT STARTS UP TOO OFTEN?

The system starts up, regulates and stops in line with the effluent input volume. Below approximately 10% of its capacity (a value set when put into service) it stops completely.
The number of start-ups is not a problem as each motor can switch on and off 150 times an hour without reaching peak current and therefore without overheating.

THE VORTEX IS SAID TO HAVE POOR HYDRAULIC OUTPUT COMPARED TO CHANNEL IMPELLERS. HOW DOES THE DIP PERFORM IN TERMS OF ENERGY CONSUMPTION?

That's correct, the DIP Vortex effect acts like an open hydraulic coupler, a "torque transmitter" which provides diphasic pumping, non-clogging, etc., but is therefore an intermediate impeller, consuming additional energy as compared to a channel impeller (20% to 25%). In the case of the DIP, however pumping from water stream intake and absorbed power adjusted at the input flow are enough in itself to compensate for this loss. Accordingly, consumption is virtually identical. This is just one example of energy savings with modulated pumping.

WHAT ARE THE RISKS OF BLOCKAGE? IS UPSTREAM SCREENING NECESSARY? ARE RAGS OR CLOTH A PROBLEM?

The DIP has absolutely nothing to fear from domestic towelettes and similar hygiene products which have become such a problem in recent years, as completely free flow sections have always been designed into the DIP system and direct in-line pumping does not include a storage area where such products are able to build up. There is no need to provide any screening systems. In addition, all DIPs have a motor torque control and automatic reverse, enabling them to tackle and reduce the volume of diapers or towelettes passing through during pumping.

In combined or rainwater networks, somtimes large unseemly objects pass through accidentally, and for this reason the DIP intake body is fitted with a "stone trap" and large, accessible service hatch.

IS THERE ANY RISK FROM ACCUMULATED GREASE OR FIBERS UPSTREAM?

No, because, when put into service, our technicians fine tune the factory settings so that the operating range matches that of the flow; then there can be no accumulations upstream, just maximum flow trace as you can see here:

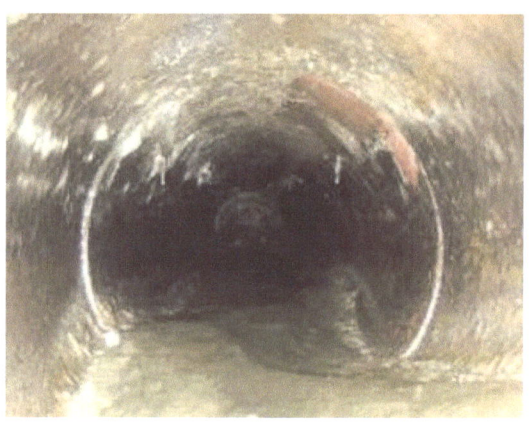

WHAT HAPPENS TO SAND AND GRIT THAT CAN COME FROM A GRAVITY NETWORK?

Sand is transported away in the flow and larger pieces of grit are caught in the rear part of the body, the "stone trap" designed for this purpose, where they can be removed via the service hatch. If the network is really full of stones and gravel, it might be useful to fit a trap in the upstream inspection chamber, for example.

WHAT HAPPENS IF THERE IS A POWER CUT?

The same as with other systems – pumping stops!
The only difference is that the back flow starts earlier
than with a submersed unit (depending on size), which
doesn't represent a great deal more security. For critical
networks, an automatic start-up generator set is the best
solution, especially as the DIP doesn't need a current
surge on start-up, the generator therefore can be sized
without incurring extra costs. Otherwise, the
maintenance-free and fuel-free solution is a safety tank,
installed upstream on same inflow.

WHAT HAPPENS IF THE MEASURING SENSOR BREAKS DOWN?

The control system ensures emergency pumping by
automatically switching the Master Pump to fallback
mode at a pre-programmed fixed speed and displays an
alarm on the screen. This degraded mode can last for up
to 150 hrs, (around one week) to allow for the changing
of the sensor.

WHAT SHOULD YOU DO IF ONE OF THE MOTORS BREAKS DOWN? WHAT ACTION SHOULD BE TAKEN?

If a fault has been identified but not resolved by the
automatic fault management system, the control system
automatically switches over to the other pump in the unit.
To remove the motor unit for service, just a few minutes
downtime is required, or even none at all if the DIP has
optional suction valves for each pump. There are only
4 to 12 nuts to remove depending on DIP size. A plate
is supplied with each system to seal the seating location
of the removed motor. Removal and replacement is
therefore straightforward and is performed in complete
safety.

HOW DO YOU KNOW THE LEVEL OF OUTLET FLOW? CAN I USE AN ELECTROMAGNETIC FLOWMETER?

You can use a classic electromagnetic spool-piece flowmeter for a DIP upstream of a treatment plant where diphasic pumping (presence of air) is not required. However, for a network unit where the injection of air is used, we suggest an adapted electromagnetic meter, not requiring straight run lengths, which provide an instant flow measurement on a signal of 4/20mA and a GPM counter.
This information can be read from the screen of the ALC control panel and sent to remote management devices. The factory will assist in sizing in accordance with the nominal discharge diameter.

SO NOW YOU HAVE IT...THE NEW TECHNOLOGY IN HANDLING WHAT YOU DON'T SEE...

AFTER THE FLUSH!

- If you are a home owner, a concerned citizen, or someone who cares about the environment, your municipality will benefit with new technology to reduce costs and relieve needs for tax increases.
- If you are a licensed operator, you could be working in a much cleaner setting, less worry about trash removal from wet wells, relief from clogged pumps, and spending your time being a better manager.
- If you are a city planner, a board member, a utility manager, and / or have other financial responsibilities to your city and its citizens, your life could be less stressful with the **DIP System®**.

IF YOU *WERE* TO FLUSH A PAIR OF JEANS DOWN YOUR TOILET…

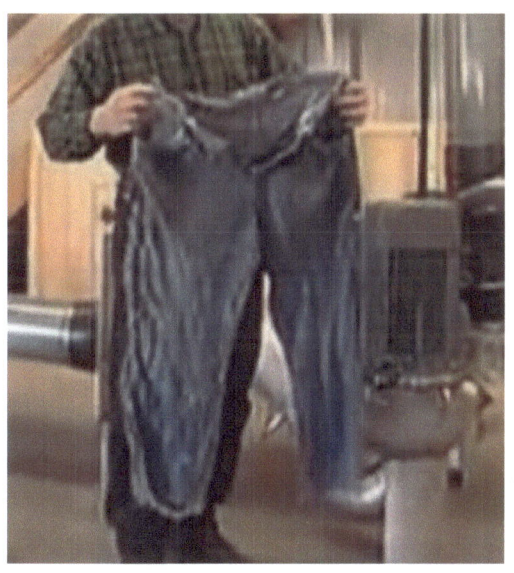

YOUR PUMPS WOULD <u>NOT</u> BE CLOGGED AND THE JEANS WOULD LOOK LIKE THIS <u>IF</u> YOU WERE USING THE DIP System® WITH THE DIPCut®

59

HERE ARE SIX DIAPERS FLUSHED AT ONCE...

AND THE DIP System® KEEPS ON PUMPING WITH NO LOSS OF CAPACITY

WANT MORE PROOF? CHECK THE LINKS BELOW:

http://youtu.be/F2nPwW3ot2Q Features of Direct in-Line Pump

http://youtu.be/hi6_hPfLdVA Controls & Operation

http://youtu.be/nFrULokUHh4 Pump body construction

http://youtu.be/TrdbMB5QEAs Shredding example

http://youtu.be/_aYEBvXxfxl No cavitation with air

http://youtu.be/HMUdBLCNP3k Diapers won't clog your lift station

http://youtu.be/JkDX01HCtnc Passing solids (tennis balls) easily

Contacts

The DIP System® is patented, and proven and widely used in Europe. There are over 1,100 systems in use in France alone. As our U.S. infrastructure is aging and in need of replacement or upgrading why should it be replaced with the same type of systems that have been in use for all these many years? We've come a long way from ancient Rome (we only added pumps and treatment plants). We can now take the next step toward the future and use the **DIP System**® technology that is now available.

The **DIP System**® is manufactured in France, and is assembled in Kansas and is shipped from Kansas.

FOR ANY ADDITIONAL INFORMATION OR TO RECEIVE A PROPOSAL IN ADDING OR ACQUIRING THIS NEW, PROVEN, TECHNOLOGY FOR YOUR WASTEWATER APPLICATION, CONTACT:

www.cbeuptime.com

C&B Equipment
9900 Pflumm, #67
Lenexa, KS 66215
(800) 475-0101 (913) 438-1212
Fax (913) 438-2221
JDunham@cbeuptime.com

About the Author

I have been fortunate to earn and enjoy a classical marketing and management career that has included key positions with well-known firms and brands. My career track has been professional growth through such positions as Product Manager, Product Marketing Manager, District Manager, Regional Manager, Sales Manager, and Vice President.

I authored industry specific books, available on Amazon.com, and published handbooks containing product information, that when applied, strengthened my customers' presence in their market segment. Production of technical installation and service manuals sharpened my writing skills. I also was recipient of national industry advertising awards. A personal relationship along with training, by the authors and founders of *"Consultative Selling"* gave me the know how to partner with, and implement the best practices for my customers.

I am skilled in all areas of executive management, including developing operations throughout the United States and Canada, building premier distribution networks, creating innovative advertising, and expanding into new industries and markets. My richly varied experiences have consistently helped my associates and customers exceed their companies' goals to achieve gains in knowledge, revenues and profits.

I have used my expertise to support customers in profit improvement or savings through offering products and services that are beneficial to their business objectives.

How can I help you?

www.ingramcontent.com/pod-product-compliance
Lightning Source LLC
Chambersburg PA
CBHW040846180526
45159CB00001B/334